What Do Animals Need?

by Margaret McNamara

Table of Contents

Words to Think About	2
Introduction	4
Chapter 1 Animals Need Food and Water	6
Chapter 2 Animals Need Oxygen	8
Chapter 3 Animals Need Shelter	10
Chapter 4 Animals Need Space	12
Conclusion	14
Glossary and Index	16

Words to Think About

energy

This cheetah uses energy to run.

grow

A baby elephant will grow to become an adult.

need

These warthogs need water to live.

shelter

This raccoon uses the tree for shelter.

space

These giraffes have space, or room, to run.

survive

This frog eats insects to survive, or stay alive.

Introduction

Animals **need** many things to **survive**, or stay alive. What do you think happens to animals that do not get the things they need?

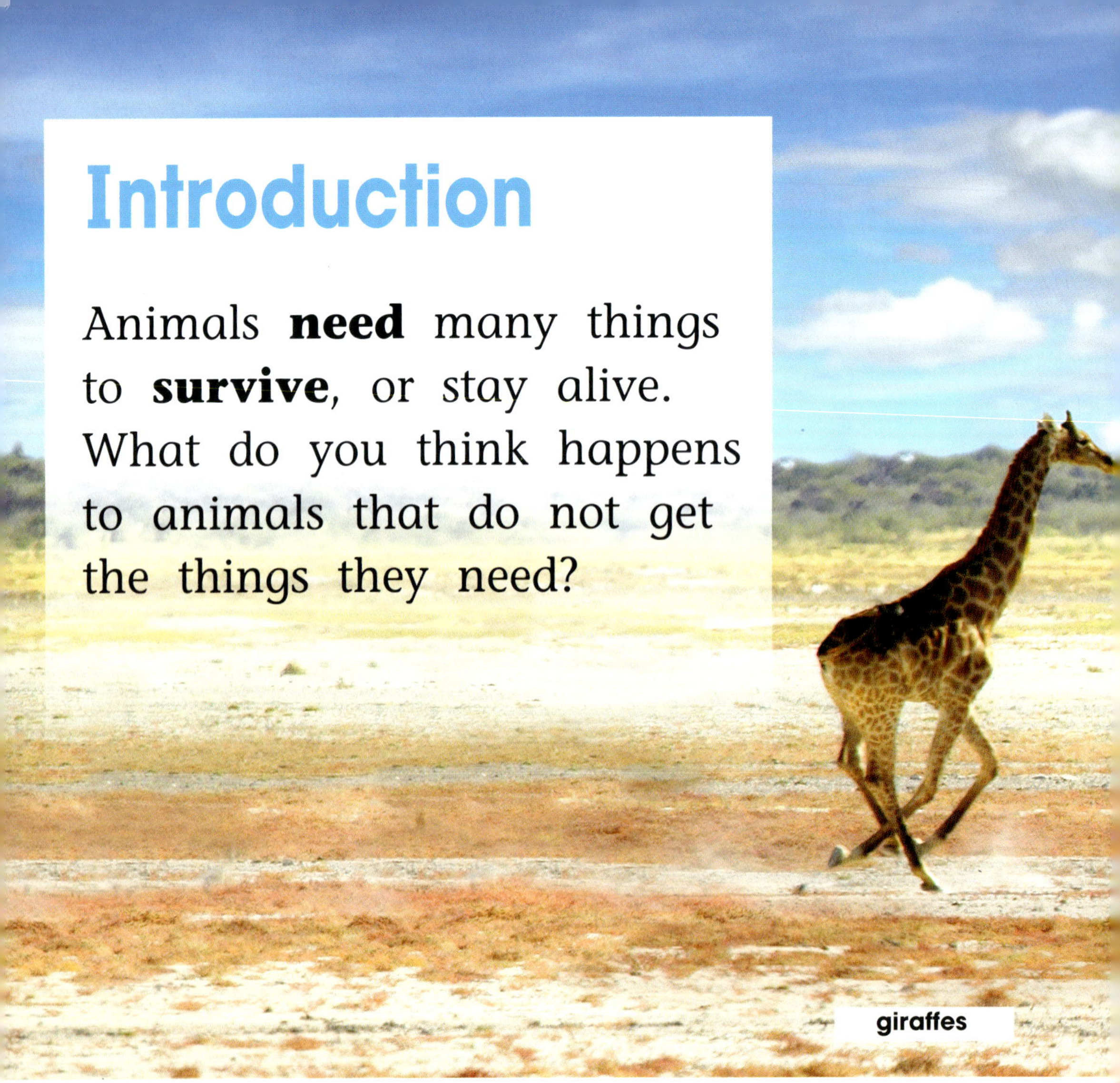

giraffes

zebra

leopard

Animals Need Food and Water

All animals—including you—need food and water to **grow**. Animals get **energy** from the food they eat. Most of an animal's body is water.

▲ This hawk uses its talons to catch fish.

In order to stay alive, animals find ways to get the food and water they need.

▲ This tiger uses its sharp teeth to get food.

▲ This elephant eats leaves to get energy.

Think About It

Herbivores are animals that eat plants.

Carnivores are animals that eat other animals.

Omnivores eat both plants and animals.

Are you an herbivore, carnivore, or omnivore?

▲ This deer drinks water to survive.

Animals Need Oxygen

Take a deep breath of air. You just took oxygen into your body. Living things use oxygen to get energy and to survive.

▲ This coral trout gets oxygen from the water. This fish uses its gills.

Some animals get oxygen from the air. Some animals that live in water get oxygen from the water.

▲ These bison breathe oxygen from the air.

▲ This orca breathes through its blowhole.

Chapter 3

Animals Need Shelter

Animals need **shelter** just like you do. Some animals find shelter. Other animals build shelter. A shelter is a safe place for an animal to live.

▲ This ground agama uses an underground shelter for protection.

If the weather is too hot or cold, then animals can protect themselves in their shelters.

▲ This raccoon finds shelter in a tree.

▲ This robin used sticks, mud, and grass to build a nest.

LOOK AT TEXT STRUCTURE

Cause and Effect

Look at the word "if." This word helps us understand the cause. What word helps us understand the effect?

▲ This clown fish finds shelter in an anemone.

Chapter 4

Animals Need Space

Animals need **space**, or room, in which to move around. If animals do not have space, then they might not find enough food and water.

▲ Ants need space, too. Ants find space in the soil.

Animals protect their space. What do you do when someone gets too close to you? How can animals find the space they need?

▲ These wild horses like to run. They need a lot of open space.

▲ These apes make loud noises to protect their space.

Conclusion

In order to survive, animals need five things. Animals need water. Food and oxygen give animals energy.

Animals need five

food

water

oxygen

Animals need shelter to stay safe. They also need space. If an animal does not have these five things, what will happen?

things to survive.

shelter

space

Glossary

energy ability to do work
See page 6.

grow to become larger and stronger
See page 6.

need must have
See page 4.

shelter a safe place
See page 10.

space room to move around
See page 12.

survive to stay alive
See page 4.

Index

animals, 4, 6–7, 9–15
energy, 6, 8, 14
food, 6–7, 12, 14

grow, 6
need, 4, 6–7, 10, 12–15
shelter, 10, 15

space, 12–13, 15
survive, 4, 8, 14
water, 6–7, 9, 12, 14

16